ENTERPRISE

EVERYTHING TO KNOW PLUS THE TEST SHUTTLES

LANE HERMANN

Get Your Printed Copy For Your Coffee Table Today!

www.createspace.com/4436848

ENTERPRISE

ISBN-13: 978-1492383079

ISBN-10: 1492383074

Copyright © 2013 by Lane Hermann

Licensed by BRISK PRODUCTIONS

All rights reserved. No part of this publication may be reproduced or transmitted in any form or by any means, electronic or mechanical, including photocopy, recording or any information storage and retrieval system, without the prior written permission of the author.

Chapters

1. Enterprise - Roots
2. Test Shuttles
3. MPTA-098
4. STA-099
5. STA-100
6. ALT-101
7. **SCA:** Shuttle Carrier Aircraft
8. **MDD**: Mate/Demate Device
9. **First Flight,** ALT: Approach and Landing Test program
10. ALT Crew Members
11. CANCELED
12. Life in Retirement
13. The Lost Years
14. Still Useful
15. Finally on put on display
16. A New Home
17. Where to see a Space Shuttle Today

BONUS - The Downey Shuttle Story, the Original 1972 Space Shuttle Mockup

ENTERPRISE intentionally set on nose during weight testing

Enterprise

Occasionally referred to as the first Space Shuttle.

To know Shuttle Enterprise we need to know her roots.

The several test aircraft leading up to the final design of the space shuttles had started in the mid 1950s

The earliest concept studies were conducted even before manned rocket space flight.

Experimental testing for the space shuttle program continued through 2005.

In 1969 Congress approved a program called the **S**pace **T**ransportation **S**ystem, or **STS**,

later to become known as The Space Shuttle.

Test Shuttles

On July 26, 1972 Congress approved $6.5 billion dollars for the Shuttle program.

Contracts were awarded for 3 Shuttle test beds, and 2 prototypes.

Project names were only used as identifiers.

No vehicle names had been chosen for any of the structures at this time.

MPTA-098   STA-099   STA-100   ALT-101   OV-102

MPTA-098

In 1974 construction started on the test bed (**MPTA**-098) The **M**ain **P**ropulsion **T**est **A**rticle

MPTA - 98 consisted of the internal structure bed of a Space Shuttle orbiter and the main engine aft-fuselage.

This included all main propulsion system plumbing and the associated electrical systems.

This structure was never given a name other than 098.

Added to this 098 set up was the MPTA-**ET** - 098, **E**xternal **T**ank.

This was the first fully functional external tank built for the Space shuttle program.

This held the Martin Marietta manufacturing Serial number, 001.

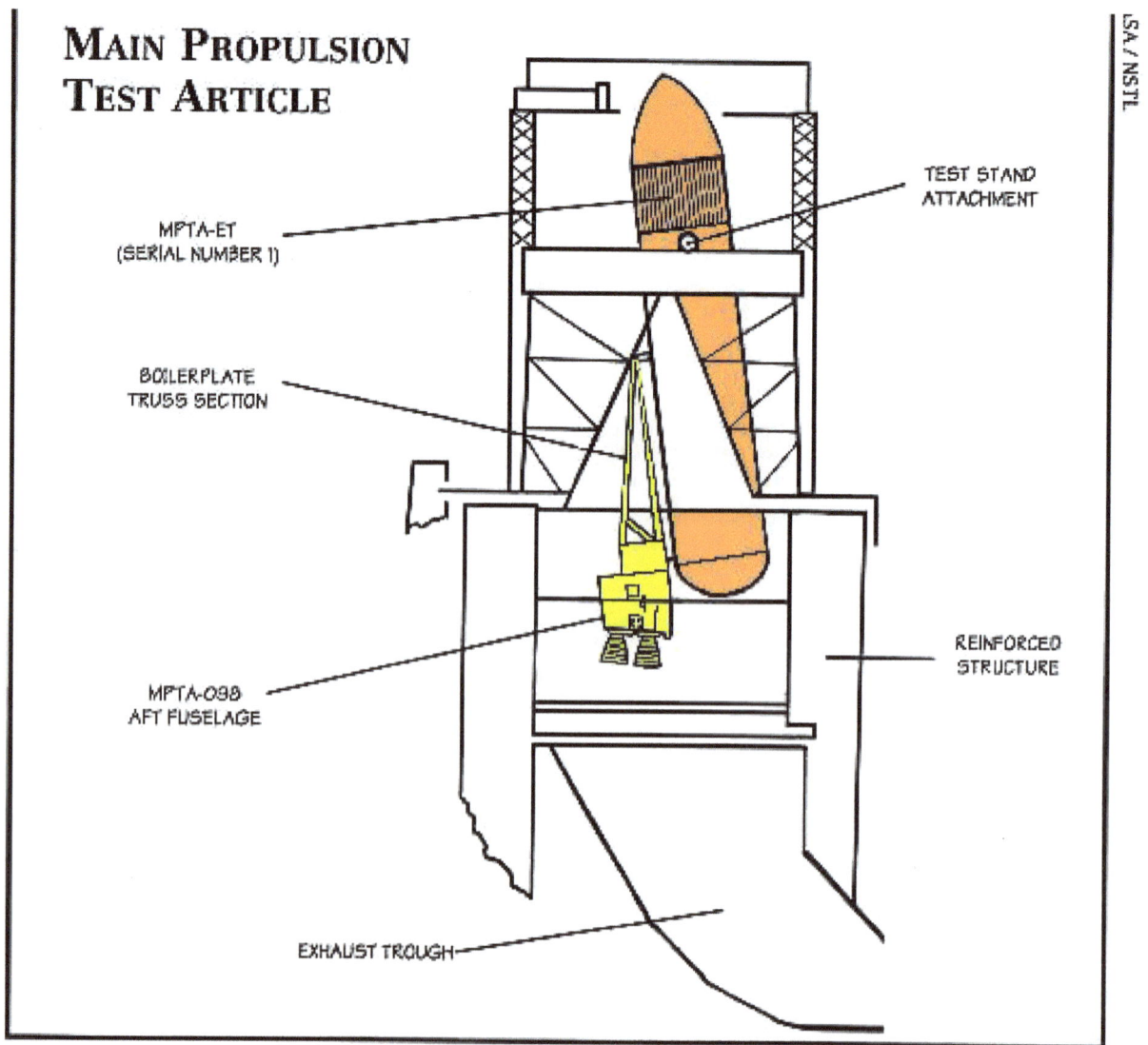

On July 2, 1979, MPTA-098 suffered major structural damage due to a fractured fuel valve allowing hydrogen to leak into the enclosed aft compartment,

Raising the pressure to beyond the structural capability of the heat shield supports.

The structure was repaired and testing continued.

With several more setbacks the Space Shuttles Main Engines were eventually perfected.

From 1981 until 1988, the MPTA-098 and MPTA-ET, external tank remained on the test stand, unused.

In late-1988, the thrust structure of 098 as was re-used.

It was the base structure for an engineering development, full size model of the proposed **Shuttle - C** launch vehicle.

The model was used by NASA and Boeing at the **K**ennedy **S**pace **C**enter (**KSC**) and the **M**arshall **S**pace **F**light **C**enter (**MSFC**)

The United States Congress in the 1990 canceled The **Shuttle - C** program and the model was disassembled.

The Main Propulsion Test Article, MPTA-098 without its truss work, is still at NASA's Stennis Space Center (**SSC**) in Mississippi today.

STA - 099

In 1975 full construction of **S**tructural **T**est **A**rticle 099 started.

STA 99 was a unique high fidelity airframe evolved under weight-saving pressures and able to handle significant structural stress.

This optimized design, it was difficult to accurately predict with the computer software available at the time. The only safe approach was to submit to intensive testing and stress analysis.

11 months of intensive vibration testing in a 43 ton steel rig.

The rig consisted of 256 hydraulic jacks, distributed over 836 load application points.

Three 1 million pound-force hydraulic cylinders were used to simulate the thrust from the Space Shuttle Main Engines.

Heating and thermal simulations were also done on the structure.

In 1978 President Jimmy Carter approved construction of four space shuttle orbiters along with structural spares for a fifth.

Later to be named

Columbia   Challenger   Discovery   Atlantis

On January 29th, 1979 NASA ordered the test vehicle **(STA-099)** converted into a space-rated **O**rbital **V**ehicle.

Space Shuttle Challenger (OV-099)

STA-100

Static Test Article 100 was built in 1977 at the **Marshall Space Flight Center (MSFC)**

It was a steel structure roughly the size, weight and shape of an empty 75 ton orbiter.

STA-100 was used as a stand in allowing facilities to be tested without requiring the use of a rare, more delicate and expensive Space Shuttle.

Moved quietly around the country to several NASA facilities for testing, STA-100 eventually ended up retired in storage at the Kennedy Space Center.

After, STA-100 had been largely forgotten in storage, a group of Japanese businessmen offered to spend $1,000,000 to modify the vehicle.

After refurbishing the steel mock-up, to more closely resemble an actual Space Shuttle,

It was named "Pathfinder" by persons unknown.

Pathfinder was displayed at the "Great Space Shuttle Exposition" in Tokyo from June 1983 to August 1984.

Eventually Pathfinder was returned to NASA and ended up at its birthplace, the Marshall Space Flight Center.

NASA did not realize the un-inventoried test shuttle Pathfinder, was actually STA-100.

It was mated to the Main Propulsion Test Article External Tank, MPTA- ET 098.

Pathfinder and MPTA- ET 098 were then put on display named generically and incorrectly after the test external tank.

Honorary **O**rbital **V**ehicle OV-098   Not STA-100.

This violated NASA's own shuttle inventory policy, as the OV- **O**rbital **V**ehicle designation was for intended space vehicles only. *All Honorary OV Shuttles are mock-ups.*

There is an urban myth which claims that non-space vehicles are all numbered with a 3 digit identifier beginning with "0". While, all space vehicles are numbered with a 3 digit identifier beginning with "1".

This myth is debunked through the example of the Space Shuttle Challenger (OV-099), which was formerly (STA-099). Clearly the Challenger was a space vehicle, and formerly a test vehicle (non-space)

The Shuttle inventory numbering system has remained consistent and has not deviated with vehicle changes.

NASA has used a variety of acronyms to describe vehicles and various changes to the vehicle's status.

This largely unknown shuttle is sometimes referred to as the 8th Shuttle.

In June of 2008 the floor on the starboard side equipment bay separated from the rest of the shuttle body.

The structural integrity of Pathfinder had deteriorated due to years of weather damage.

It has since been restored and is on display at the Marshall Space Flight Center today.

ALT-101

Known sometimes as the First Space shuttle.

In 1974 work started on the first prototype Shuttle originally named **Constitution ALT-101**

ATL stands for the **A**pproach and **L**anding **T**est program.

Coming close to completion it was decided that both prototype shuttles 101 and 102 would fly missions in to space.

Constitution ALT-101 was now Constitution Orbital Vehicle 101.

The American viewers of the popular TV Science Fiction show Star Trek started a write-in campaign urging the White House to select the name Enterprise for OV-101.

President Gerald Ford approved the name change with an executive order.

The American manned space plane **Enterprise OV-101** was rolled out to the public on September 17, 1976.

At this time Enterprise weighing in at 150,000 pounds was only configured to be the world's largest glider and was not even close to being a Space rated vehicle.

SCA
Shuttle Carrier Aircraft

Before Enterprise could fly, the engineless shuttle had to hitch a piggy back ride.

The Lockheed C-5 Galaxy was considered for the shuttle-carrier role, but was rejected.

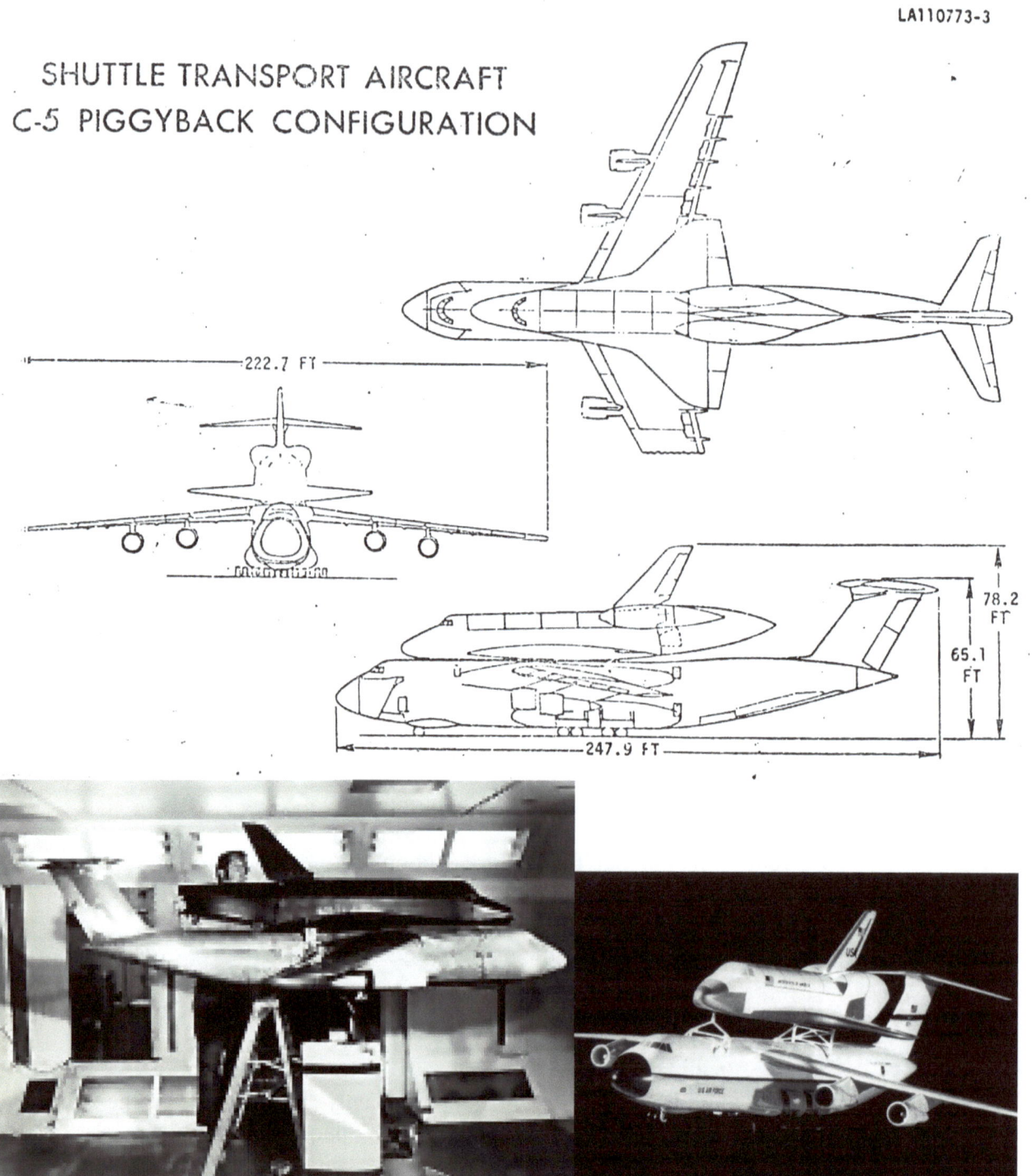

The C-5's high-wing design was problematic.

Attempting to address both wing and weight issues,

Both Rockwell and Lockheed considered 2 C-5s melded together.

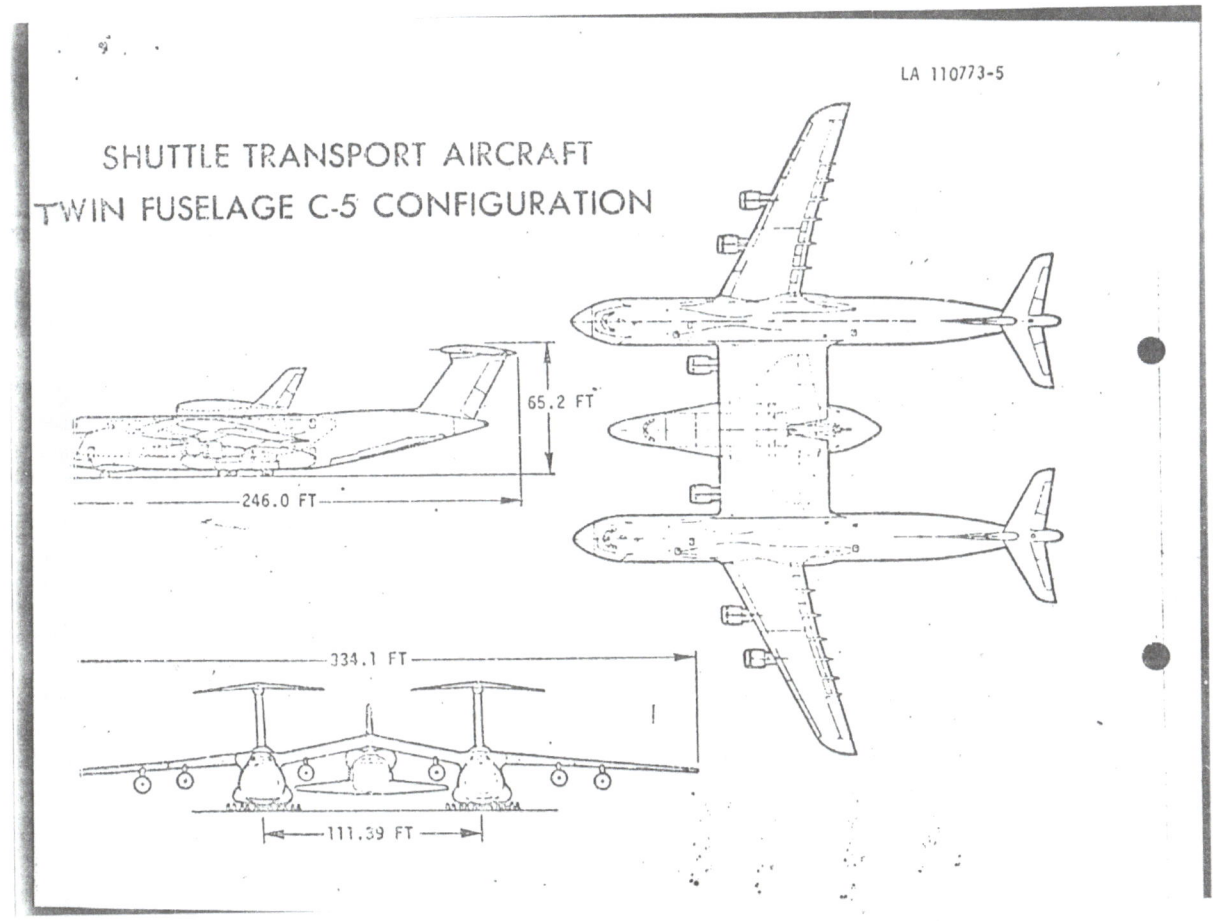

Also the U.S. Air Force would have retained ownership of the C-5.

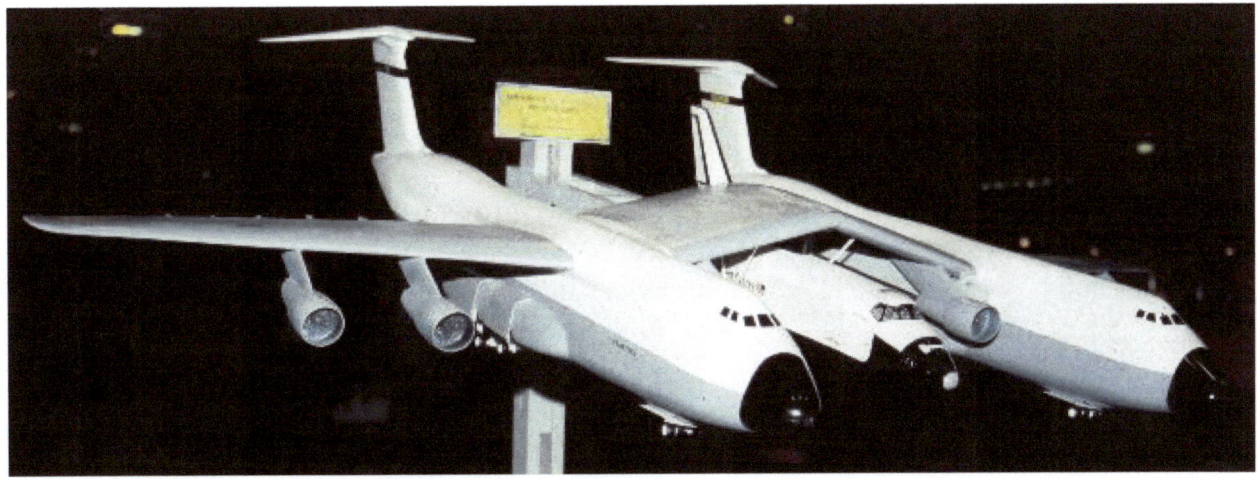

The 747's low-wing design was more favorable; while NASA would be able own the 747s outright.

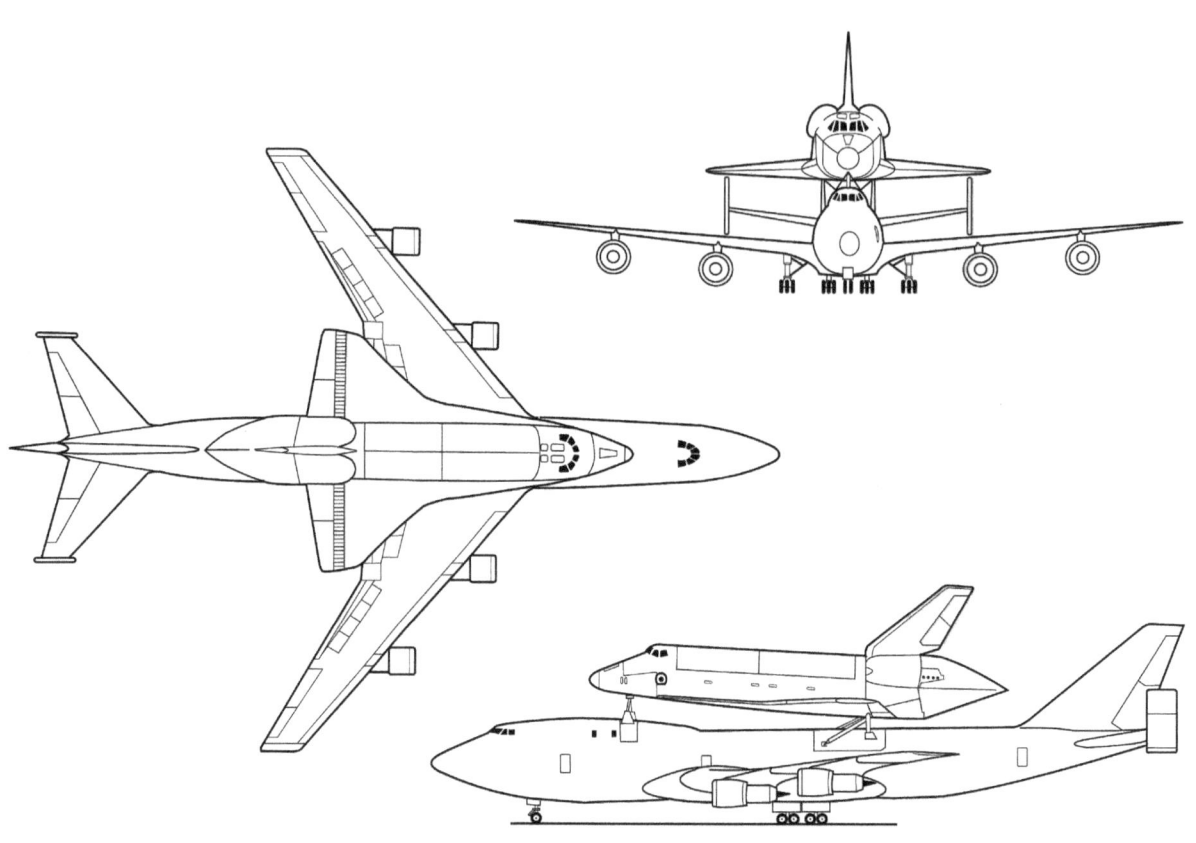

 Dryden Flight Research Center February 1998
Space Shuttle mated to 747 Shuttle Carrier Aircraft (SCA) 3-view

In 1974, NASA purchased a Boeing 747-100 series from American Airlines.

Registered as N905NA, Known as SCA 905.

In 1990 NASA procured a second surplus 747-100SR from Japan Airlines.

Registered N911NA, SCA 119 entered service after undergoing modifications similar to N905NA.

It was first used in 1991

After a series of wake vortex research flights,

747 and both C-5 Models with the orbiter were studied in wind tunnels.

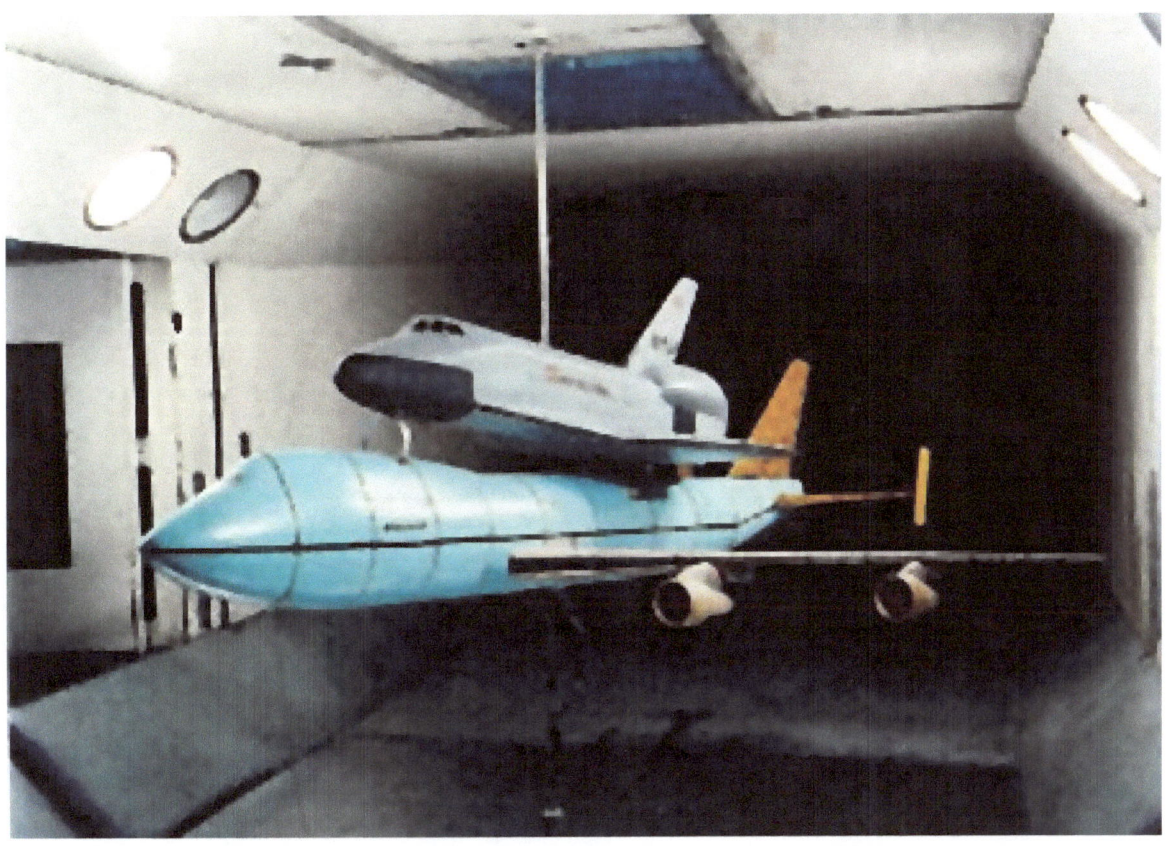

In preparation for the ATL program, the SCA 905 was sent to Boeing for modification in 1976.

The First-class seats were kept for NASA passengers.

The passenger seats were taken out.

Its main cabin and insulation were stripped.

1,400 lbs of iron and 7,000 lbs of gravel ballast were added.

Ballast at the front of the 747 counteracts the rear-weight bias of the shuttle.

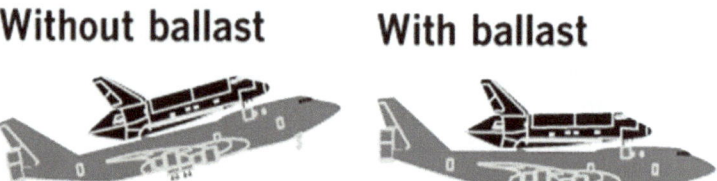

Mounting struts added, and the fuselage strengthened.

Vertical stabilizers were added to the tail to aid with stability when the Orbiter was being carried.

The avionics and engines were also upgraded.

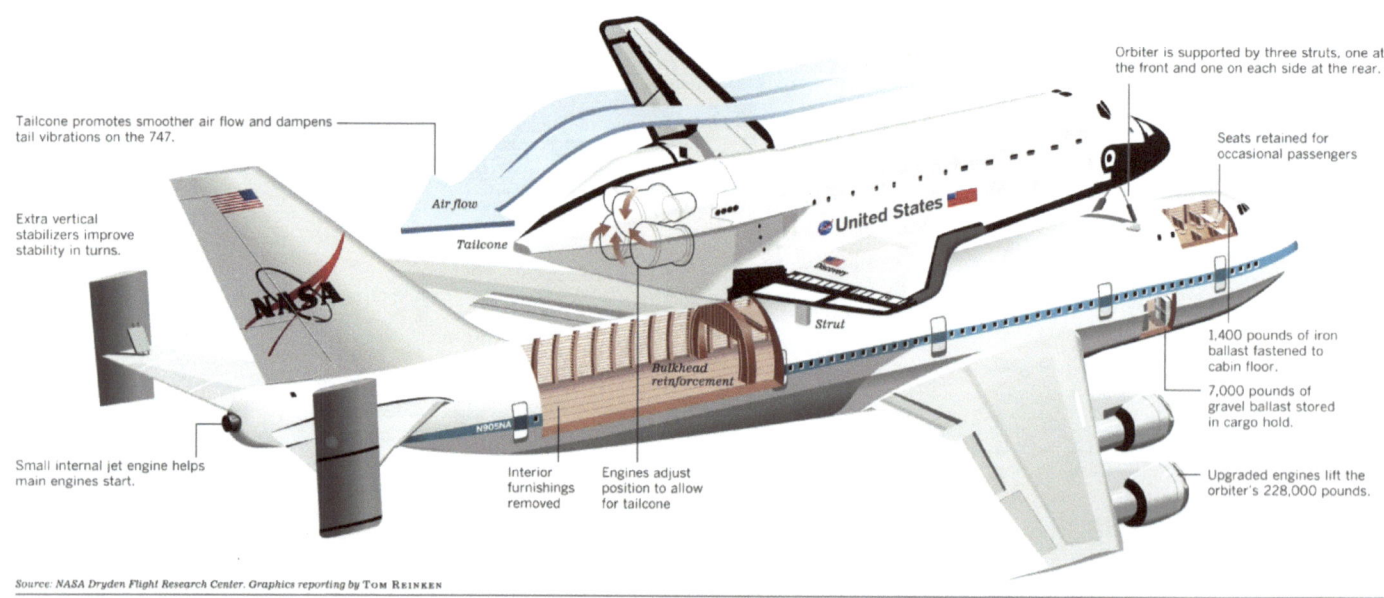

A flight crew escape system was added.

An exit tunnel extending from the flight deck to a hatch in the bottom of the fuselage, including pyrotechnics for the hatch release was added.

Falling down the slide and out of the bottom the crew could get to safety.

Any passengers would be able to escape through the windows that were wired to be blown out with pyrotechnics.

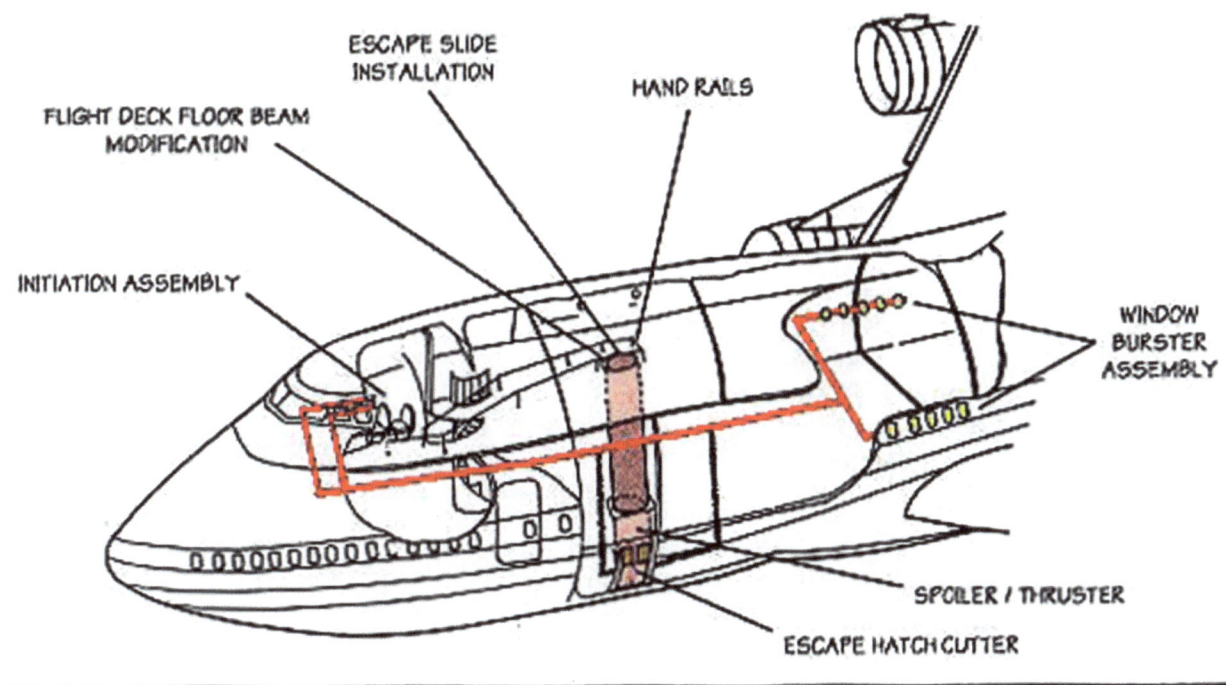

The flight crew escape tunnel system was later removed following the completion of the Approach and Landing Tests.

General characteristics

- **Crew:** 4: pilot, co-pilot, 2 flight engineers (1 flight engineer when not carrying Shuttle)
- **Length:** 231 ft 4 in (70.5 m)
- **Wingspan:** 195 ft 8 in (59.7 m)
- **Height:** 63 ft 5 in (19.3 m)
- **Wing area:** 5,500 ft² (510 m²)
- **Vertical tip fins on horizontal stabilizers:** 20 ft. 10 in. high, 9 ft. 7 in. long
- **Empty weight:** SCA 905 318,000 lbs (144,200 kg) SCA 119, 323,034 lbs
- **Max. takeoff weight:** 710,000 lb (322,000 kg)
- **Power plant:** 4 × Pratt & Whitney JT9D-7J turbofans, 50,000 lbs (222 kN) of thrust each

Performance

- **Aviation Fuel capacity:** 47,210 gallons (316,307 lbs)
- **Cruise speed:** Mach 0.6 (397 knots, 457 mph, 735 km/h)
- **Range:** 1,150 Miles (1,000 Nautical Miles, 1,850 km) while carrying Shuttle.

   Without a shuttle Maximum range 5,500 Nautical Miles

- **Service ceiling altitude:** 13,000 - 15,000 ft (with Shuttle)

   Without a shuttle 24,000 -26,000 ft

MDD

The MDD: **M**ate/**D**emate **D**evice was completed in 1976.

The MDD consists of two 100-foot tall gantry towers.

Three cranes 70 feet out from the main tower units would guide and control a large lift beam cradle that attaches to the orbiters to raise and lower them on to the SCA.

❶ Shuttle is hoisted and receives a visual inspection. Residual fuel is flushed from engine valves and plumbing; other toxic substances removed.

❷ Technicians secure aerodynamic 10,000-pound aluminum tail cone that helps eliminate drag during flight back to Florida.

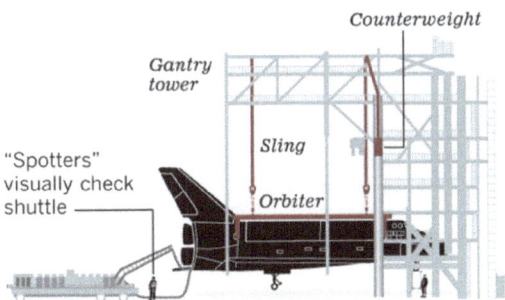

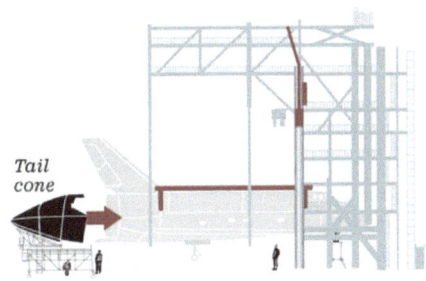

❸ Engine nozzles and elevons are locked in place. Shuttle is raised 60 feet to accommodate mounting on 747 shuttle carrier.

❹ The 747 is towed into place and the shuttle is attached to three struts that match fittings on the external fuel tank used at launch.

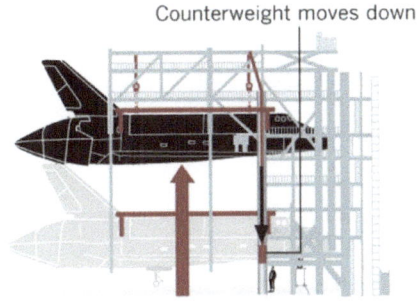

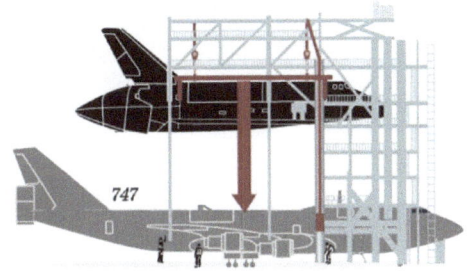

On February 7th 1977 MDD preparations with Enterprise started and were completed on February 15th.

During this time a technical problem arose when SCA 905 was rolled in under the suspended Shuttle.

They had gotten to the point where they were ready to lower Enterprise onto the 747.

The orbiter didn't fit!

The forward nose strut on the 747 had to be moved.

The operation lasted almost 14 hours.

First Flight

On January 31, 1977, Enterprise left the Rockwell Palmdale plant 42.

Enterprise was towed for 36 mile overland to NASA's Dryden Flight Research Facility at Edwards Air Force Base.

Officially entering (ALT),

The **A**pproach and **L**anding **T**est program for the next nine months.

The ALT program involved ground tests and flight tests.

The ground tests included taxi tests of the 747 Shuttle Carrier Aircraft with the Enterprise mated on top.

This was to determine structural loads and responses and assess the capability in ground handling and control characteristics up to flight takeoff speed.

The taxi tests also validated 747 steering and braking with the orbiter attached.

Five captive unmanned flights of the Enterprise mounted atop the 747 with the systems inert were conducted to assess the structural integrity and performance handling qualities of the mated aircraft.

These tests were followed with three test flights with *Enterprise* manned to test the shuttle flight control systems while still attached to the SCA.

On August 12, 1977, the space shuttle *Enterprise* separated from the SCA flew on its own for the first time.

This was the first of 5 free flights with Enterprise separating and landing on its own.

The last 2 of these free flights would be flown with the tailcone removed.

Dryden Flight Research Center ECN 8923 Photographed 10/77
Shuttle Enterprise separates from the 747 for its first tailcone-off flight.

The fifth landing was the first on a concrete runway.

During the landing a serious flight control system problem occurred which generated uncontrolled oscillations.

Causing the Shuttle to pitch, roll, and skip down the runway.

Broadcast live on TV and in front of hundreds of people along the runway.

This was initially reported as intentionally done by the pilot. It was not.

The problem was immediately investigated and fixed in the computerized guidance digital fly by wire system.

In 1972 a NASA F-8 was the first aircraft ever equipped with the Digital Fly-By-Wire flight control system.

This aided engineers in developing a safer, digital fly-by-wire software control system for the Space Shuttle as well as other aircraft.

When the ALT program ended, the Shuttle nose strut on the SCA was lowered by 6 degrees.

4 more Ferry flights were conducted before Enterprise was modified for new ground vibration tests.

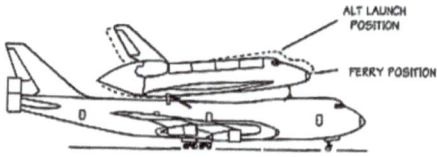

On March 13th 1978 Enterprise was delivered to the Marshal Flight Center.

From there it was mounted vertically to MPTA- ET 098 at **S**tennis **S**pace **C**enter (SCC).

Enterprise would spend over the next year in the tower experiencing vertical vibration tests.

Shuttle Approach and Landing Tests
Flight Date Description of Flight and Crew Members

Flight 1 Feb 18, 1977, Captive Inert
Unmanned inert Orbiter (Enterprise) mated to Shuttle Carrier Aircraft (SCA) to evaluate low speed performance and handling qualities of Orbiter/SCA combination.
Flight Time: 2 hours 10 minutes.
SCA Crew: Fitzhugh L. Fulton, Jr., Thomas C. McMurtry, Vic Horton, and Skip Guidry.

Flight 2 Feb 22, 1977, Captive Inert
Unmanned inert Orbiter (Enterprise) mated to SCA to demonstrate flutter free envelope. Flight Time: 3 hours 15 minutes.
SCA Crew: Fitzhugh L. Fulton, Jr., Thomas C. McMurtry, Vic Horton, and Skip Guidry.

Flight 3 Feb 25, 1977, Captive Inert
Unmanned inert Orbiter (Enterprise) mated to SCA to complete flutter and stability testing. Flight Time: 2 hours 30 minutes.
SCA Crew: Fitzhugh L. Fulton, Jr., Thomas C. McMurtry, Vic Horton, and Skip Guidry.

Flight 4 Feb 28, 1977, Captive Inert
Unmanned inert Orbiter (Enterprise) mated to SCA to evaluate configuration variables. Flight Time: 2 hours 11 minutes.
SCA Crew: FitzhughL. Fulton, Jr., Thomas C. McMurtry, Vic Horton, and Skip Guidry.

Flight 5 Mar 2, 1977, Captive Inert
Unmanned inert Orbiter (Enterprise) mated to SCA to evaluate maneuver performance and procedures.
Flight Time: 1 hour 40 minutes.
SCA Crew: Fitzhugh L. Fulton, Jr., A. J.Roy, Vic Horton, and Skip Guidry.

Flight 1A Jun 18, 1977, Captive Active
First manned captive active flight Manned active Orbiter (Enterprise) mated to SCA for initial performance checks of Orbiter Flight Control System. Flight Time: 56 minutes.
SCA Crew: Fitzhugh L. Fulton, Jr., Thomas C. McMurtry, Vic Horton, and Skip Guidry. Orbiter Crew: Fred W. Haise, Jr. and C. Gordon Fullerton, Jr.

Flight 1 Jun 28, 1977, Captive Active
Manned captive active flight Manned active Orbiter (Enterprise) mated to SCA to verify conditions in preparation for free flight.
Flight Time: 1 hour 3 minutes.
SCA Crew: Fitzhugh L. Fulton, Jr. and Thomas C. McMurtry. Orbiter Crew: Joe H. Engle and Richard H. Truly.

Flight 3 Jul 26, 1977, Captive Active
Manned captive active flight Manned active Orbiter (Enterprise) mated to SCA to verify conditions in preparation for free flight.
Flight Time: 59 minutes
SCA Crew: Fitzhugh L. Fulton, Jr. and Thomas C. McMurtry. Orbiter Crew: Fred W. Haise, Jr. and C. Gordon Fullerton, Jr

Free Flight 1 Aug 12, 1977
First manned free flight Manned Orbiter (Enterprise) with tailcone on, released from SCA to verify handling qualities of Orbiter.
Flight Time: 53 minutes 51 seconds
SCA Crew: Fitzhugh L. Fulton, Jr. and Thomas C. McMurtry. Orbiter Crew: Fred W. Haise, Jr. and C. Gordon Fullerton, Jr.

Free Flight 2 Sep 13, 1977
Manned free flight Manned Orbiter (Enterprise) released from SCA to verify characteristics of Orbiter.
Flight Time: 54 minutes 55 seconds
SCA Crew: Fitzhugh L. Fulton, Jr. and Thomas C. McMurtry. Orbiter Crew: Joe H. Engle and Richard H. Truly.

Free Flight 3 Sep 23, 1977
Manned free flight Manned Orbiter (Enterprise) released from SCA to evaluate Orbiter handling characteristics.
Flight Time: 51 minutes 12 seconds
SCA Crew: Fitzhugh L. Fulton, Jr. and Thomas C. McMurtry. Orbiter Crew: Fred W. Haise, Jr. and C. Gordon Fullerton, Jr.

Free Flight 4 Oct 12, 1977
Manned free flight Manned Orbiter (Enterprise) with **tailcone off** and three simulated engine bells installed, released from SCA to evaluate Orbiter handling characteristics. Flight Time: 1 hour 7 minutes 48 seconds
SCA Crew: Fitzhugh L. Fulton, Jr. and Thomas C. McMurtry. Orbiter Crew: Joe H. Engle and Richard H. Truly.

Free Flight 5 Oct 26, 1977
Manned free flight Manned Orbiter (Enterprise) with **tailcone off**, released from SCA to evaluate performance of landing gear on paved runway. Flight Time: 54 minutes 42 seconds.
SCA Crew: Fitzhugh L. Fulton, Jr. and Thomas C. McMurtry. Orbiter Crew: Fred W. Haise, Jr. and C. Gordon Fullerton, Jr

ALT Orbiter Crew:

Apollo 13 Astronaut,
Fred W. Haise, Jr.

Space Shuttle Astronaut, STS-3, STS-51F
C. Gordon Fullerton, Jr   October 11, 1936 - August 21, 2013 (aged 76)

X-15 program, Space Shuttle Astronaut, STS - 2, STS – 51-I
Joe H. Engle

Space Shuttle Astronaut, STS - 2, STS -26
Richard H. Truly

SCA ALT Crew

Flight crewmembers of Enterprise and the host NASA 747 Shuttle Carrier Aircraft include, from left, Fitz Fulton, Gordon Fullerton, Vic Horton, Fred Haise, Vincent Alvarez (flight engineer) and Tom McMurtry.

Fitzhugh L. Fulton, Jr. - Pilot Astronaut Badge/wings include: X-15, YF-12A and SR-71(YF-12C)

Vic Horton - Pilot Astronaut Badge/wings include: SR-71(YF-12C)

Thomas C. McMurtry - Pilot Astronaut Badge/wings include: SR-71(YF-12C)

Skip Guidry – No Picture available

Arda Joseph "A.J." Roy Jr., - January 2, 2008 (aged 78)

At the conclusion of this testing, Enterprise OV-101 was supposed to be taken back to Palmdale for retrofitting as a fully spaceflight capable vehicle.

Testing proved a number of significant design changes had taken place.

Enterprise would be required to be completely torn apart to be rebuilt for space flight.

Unlike the other Shuttles, Enterprise was largely made of polyurethane foam and fiberglass around an aluminum frame.

On January 29th 1979 NASA took the option to instead convert the incomplete, superior high fidelity Structural Test Article, 099, Orbiter Challenger

The 3 scheduled missions for Enterprise were canceled and Enterprise would never fly to space.

#	Date	Designation	Launch pad	Notes
1	July 16, 1981	STS-17 CANCELED	39-A	deployment of Intelsat V satellite and retrieval of Long Duration Exposure Facility
2	September 30, 1981	STS-20 CANCELED	39-A	Spacelab mission
3	November 25, 1981	STS-22 CANCELED	39-A	Spacelab mission

Test Shuttle Enterprise's cockpit

Space Shuttle Orbiter first generation cockpit

On April 10th 1979 enterprise was ferried to the Kennedy Space Center to be fitted to a full Shuttle external tank and solid rocket booster (SRB) stack on the mobile launcher platform.

Transported via the mobile launcher platform to Launch Complex 39-A.

Enterprise served as practice for launch, and complex 39-A, fit-check verifications.

Once at the pad, Enterprise helped validate launch pad procedures.

The biggest test was a "wet" Countdown Dress Rehearsal to simulate External Tank fueling operations for launch.

Hundreds of thousands of gallons of Liquid Hydrogen and Liquid Oxygen were used during this testing.

However, something quite disturbing was discovered during this test.

Ice was building up on the External Tank.

The solution would be the addition of the Gaseous Oxygen (GOX) vent arm and vent hood. (Referred to throughout the program as the "beanie cap")

Additionally there was an Ice inspection team used for every shuttle launch.

The "beanie cap" is currently on display at The Kennedy Visitors Space Center, next to Space Shuttle Atlantis.

Life in Retirement

With the completion of critical testing, Enterprise spent the month of August touring 6 American cities.

Enterprise was then returned to Rockwell's plant in Palmdale.

For the first time Enterprise was retired.

In October 1979 Enterprise was partially disassembled to allow certain components to be refurbished and reused in other shuttles.

On July 4th 1982 a stripped down Enterprise was towed out and used as a back drop for President Regan's speech while Columbia returned from Space.

Almost a year later Enterprise OV-101 was visibly restored and sent on an international public relations ambassador tour to be highlighted at the Paris Air show.
Stops included Canada, Iceland, England, France, Italy, and Germany.

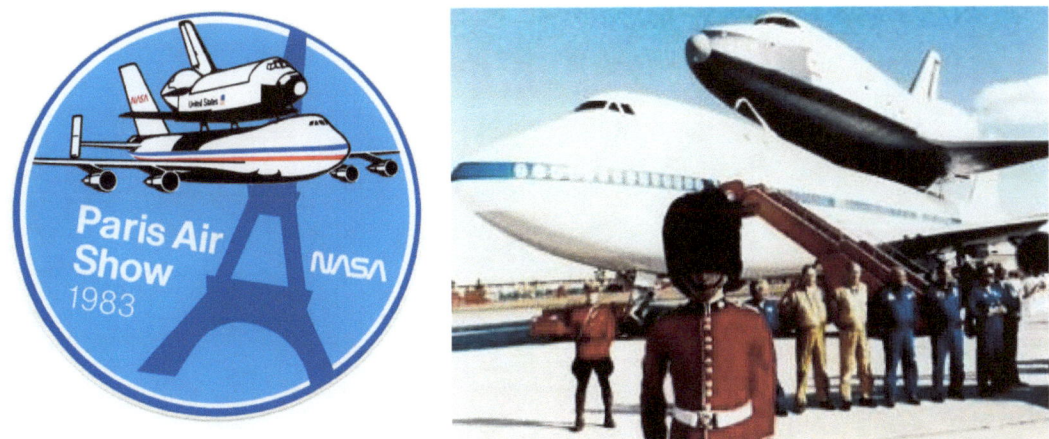

SCA 905's American red white and blue cheat line was now painted NASA blue.

Less than a year later Enterprise took the SCA ride to mobile Alabama.

Placed on a barge for the first time the Shuttle Enterprise was floated to New Orleans for the 1984 World's Fair.

After the world's fair, Enterprise was sent to Vandenberg Air Force Base and used for practice and fit-check verifications at the new **S**huttle **L**aunch **C**enter, SLC-6 complex.

The Lost Years

A year later on November 18, 1985, Enterprise was retired for the second time and sent to Dulles Airport, Washington, D.C., where it became property of the Smithsonian Institution.

Two months later, the sister ship Challenger OV-099 that took Enterprise's place in the operational Shuttle fleet was lost with all crew members aboard.

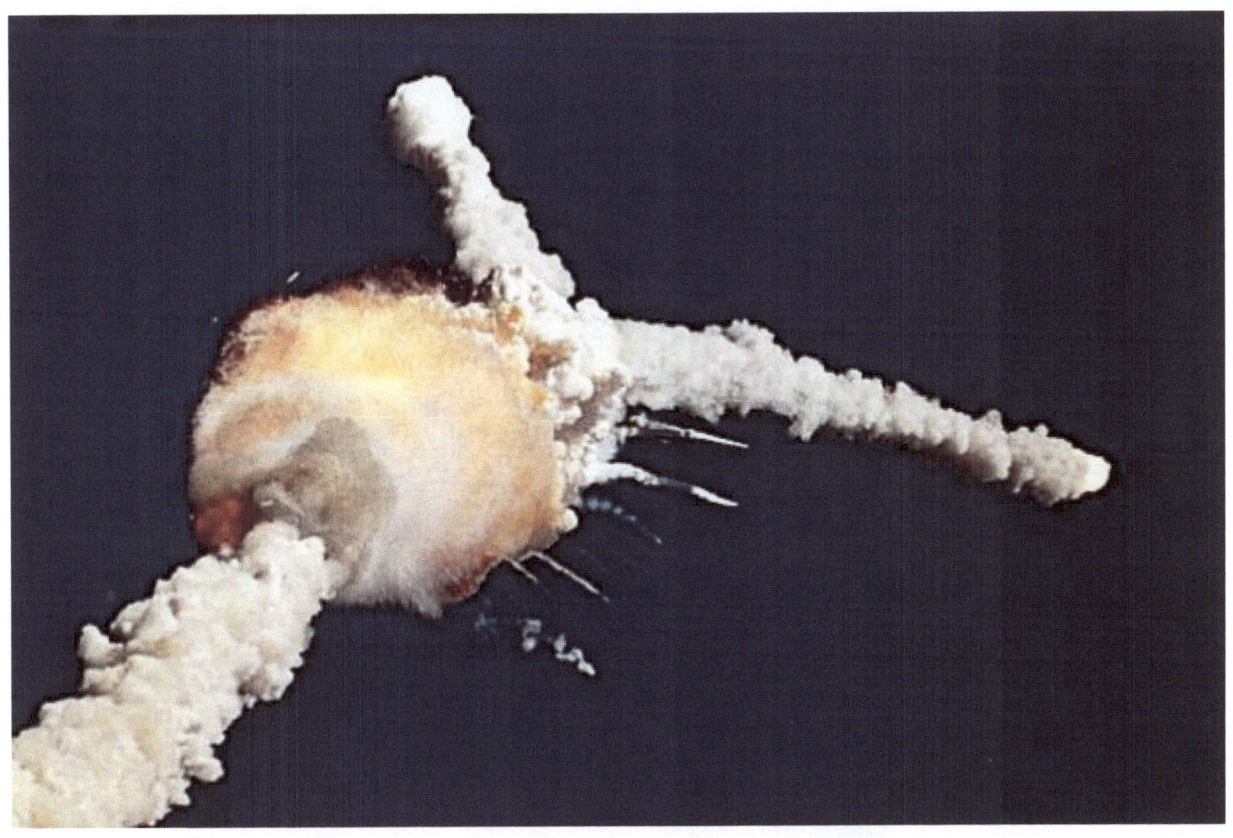

In the months that followed, NASA and the U.S. Congress agreed that a fifth Space Shuttle orbiter was needed to replace Challenger.

For a time, consideration was again given to converting Enterprise into a space-worthy Shuttle orbiter.

However, it was not to be.

With contractors able to catch up, spare parts would be used to build a completely new Shuttle orbiter.

A move that was both quicker and cheaper than converting Enterprise.

The new orbiter would be Endeavour, OV-105.

It had been determined for Enterprise to be use, the entire crew cabin would have to be cut out and replaced.

The foam body could not be converted either.

An unpressurised crew cabin, Enterprise had been stripped of any useful equipment already.

Still Useful

NASA still used Enterprise to test improvements in the wake of the loss of Challenger.

The breaks had failed on most of the 24 Space Shuttle landings to date and new safety measures were being looked at.

In 1987, test were conducted with Enterprise of a barricade landing orbiter arresting system.

Similar to that used on aircraft carriers.

The Smithsonian Institution however, due to the lack of a suitable facility left Enterprise outside, exposed to the Washington area elements for three years until a temporary hangar was ready.

In 1988 Enterprise was warehoused from the public without any environmental controls.

Enterprise was finally moved to the Space Hangar in Nov 2003 and repaired.

In 1990, the nose gear was removed for studies that ended up being deferred.

NASA continued with testing the placement of new antennas on a shuttle window using Enterprise.

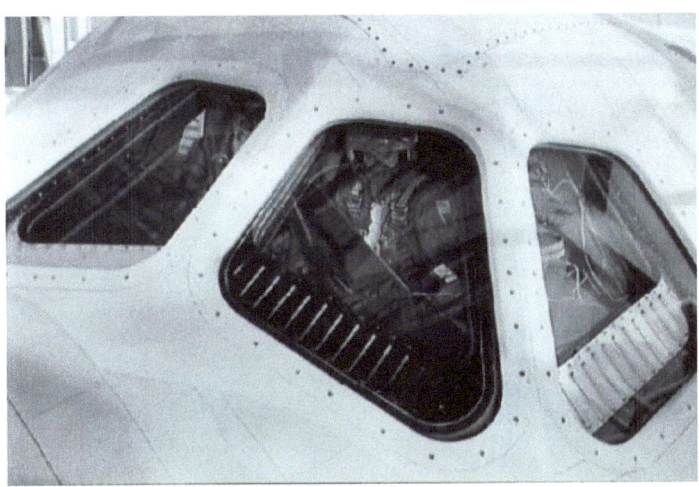

In 1996, the nose gear was finally returned to the museum.

In 1997, the landing gear was removed again for studies on an improved nose gear.

In 2003, after the breakup of Columbia during re-entry, an investigation was started.

A fiberglass panel from Enterprise's wing was removed to perform foam strike tests.

An air gun fired external tank foam insulation at nearly 530 miles per hour at the orbiter wing leading edge panel.

It was intended to simulate the foam strike that fatally wounded Columbia during launch.

The **R**einforced **C**arbon-**C**arbon (RCC) panel on Columbia had only 40% of the strength of Enterprise's stronger fiberglass panel.

This result suggested that the RCC leading edge would have been shattered.

Additional tests on the fiberglass were canceled.

A panel from *Discovery* was tested to determine the effects of the foam on a similarly-aged RCC leading edge.

On July 7, 2003, a foam impact test created a hole 16 inches by 16.7 inches in Discovery's wing.

Real shuttle tiles were attached to the left landing gear door for foam impact testing as well.

These are the only shuttle tiles ever placed on Enterprise.

Finally on put on display

After the warehouse damage to the tail and other areas were repaired.

The deteriorated paint had been stripped down, repainted, and restored.

Enterprise finally went on display in the Smithsonian National Air and Space Museum.

This would be the first public display of Enterprise in 18 years.

The ageing Space Shuttle fleet was officially retired in 2011.

It was known as early as 2010, Enterprise was being kicked out to make room for the Shuttle fleet's flagship Orbiter,

Space Shuttle Discovery.

Preparation inspections were underway for Enterprises removal.

Enterprise spent 8 years on display

Enterprise was found to be in good condition by NASA engineers.

138 "discrepancies" were noted by the NASA inspection teams:

39 instances of structural corrosion, 25 occurrences of ground damaged structures,

60 loose or missing cotter pins or safety wires, and 14 other subsystem anomalies.

Several issues remained when the updated landing gear was reinstalled in 1997.

Little care was taken on what was considered a museum piece at the time.

The landing gear had never been retracted since being reinstalled and was misaligned.

Hydraulic lines were not fully installed.

A gear door hook was put on backwards.

NASA engineers also found Enterprise's landing gear was designed differently than space flight Shuttle Orbiters

Most critical was the SCA nose attachment point had corrosion.

This had to be cut out and re-welded before Enterprise could be ferried anywhere.

The crew cabin attachment points were also badly corroded and needed repairing.

A New Home

On December 12, 2011, title ownership of the *Enterprise* was officially transferred to the Intrepid Sea, Air & Space Museum in New York City.

On April 17, 2012, Discovery arrived to take Enterprise's place.

April 27, 2012, Enterprise took off from Dulles International Airport en-route to a fly-by tour of New York and the city's famous landmarks.

Hangar 12 at JFK International Airport,

Enterprise was then transferred from the SCA to a waiting barge.

On June 3, the transport barge took Enterprise to Jersey City.

A gust of wind blew the barge towards a bridge piling.

Enterprise sustained cosmetic damage to the right "starboard" wingtip.

The damage was painted over.

June 6, Enterprise was hoisted onto the deck of the aircraft carrier Intrepid Museum.

Under a pressurized, air-supported fabric bubble Enterprise went on public display July 19, 2012.

On October 29, 2012, Hurricane Sandy caused Intrepid to flood, knocking out the museum's electrical power and both backup generators.

The loss of power caused the Space Shuttle Pavilion to deflate.

High winds from the hurricane caused the fabric of the Pavilion to tear and collapse around the orbiter.

Minor damage was first spotted on the vertical stabilizer of the orbiter, as a portion of the tail fin above the rudder/speedbrake had broken off.

There was other minor damage.

Temporary scaffolding and sheeting was erected around Enterprise to protect it from the environment during the straight forward aircraft repairs.

Fully repaired Enterprise reopened to the public on July 10, 2013.

On March 13, 2013 Shuttle Enterprise was listed on the National Register of Historic Places.

Reference number 13000071,

The historic significance criteria are in space exploration, transportation, and engineering

In Recognition of Enterprise's role in the development of the Space Shuttle Program.

Where to see a Space Shuttle Today

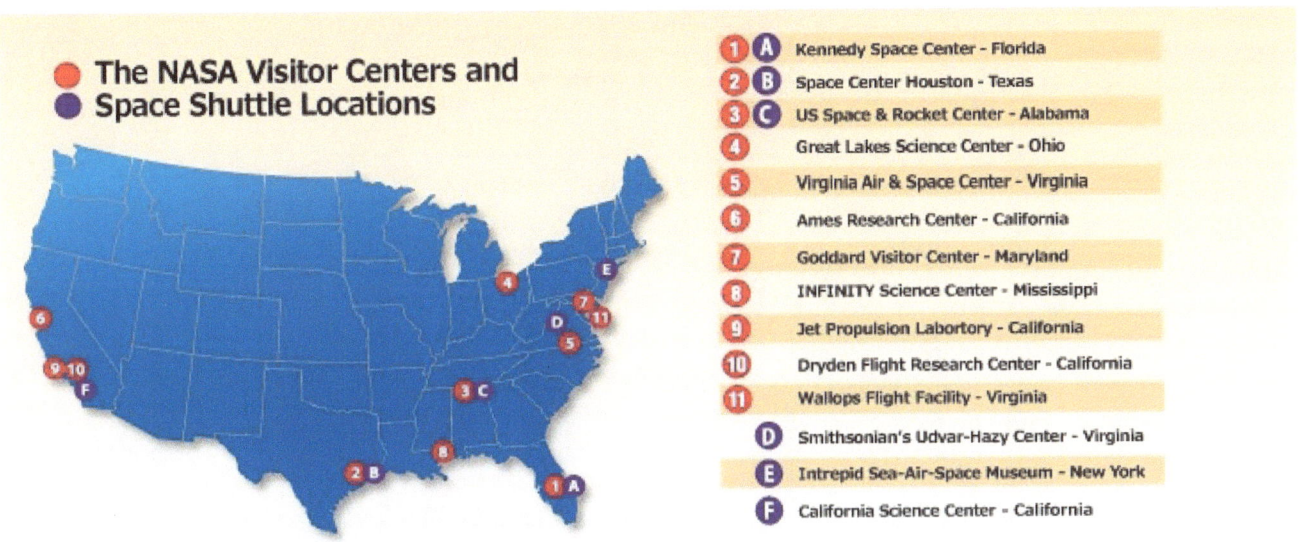

A. Kennedy Space Center Visitor Complex for the Kennedy Space Center (KSC)
<u>2 Shuttles are available to be seen.</u>

Space Shuttle Atlantis (OV-104) and Full-size **Mock up** Shuttle *Inspiration*
Shuttle Inspiration has been in place for several decades.

Full-size **Mock up** Shuttle *Inspiration,* Shuttle to Tomorrow – operated as part of the Kennedy Space Center Visitor Complex

B. Space Center Houston, the visitor center for NASA's Johnson Space Center (**JSC**)
Full-size **Mock up** Shuttle **Explorer** comes to Houston from **KSC** and will be displayed on top of **Shuttle** Carrier Aircraft **SCA 905**

Mock up Shuttle **Explorer** has been incorrectly named (honorary Orbiter Vehicle Designation: (OV-100)
Space Center Houston has also started a contest to re-name **Explorer**

C. The U.S. Space and Rocket Center for NASA's Marshall Space Flight Center (**MSFC**) in Huntsville Alabama.

Shuttle Pathfinder (STA-100) and **MPTA-ET - 098**, **E**xternal **Ta**nk
Static **T**est **A**rticle 100 **(STA-100)** sits on top of External Tank -098

Shuttle Pathfinder (STA-100) has been incorrectly named (honorary Orbiter Vehicle Designation: (OV-098)

D. The Smithsonian's Steven F. Udvar-Hazy Center of the National Air and Space Museum Chantilly, Va.

Residing in the James S. McDonnell Space Hanger, wing of the museum,

Space Shuttle **Discovery (OV-103)**

On April 19, 2012 **Space Shuttle Discovery (OV-103)** officially replaced **Shuttle Enterprise (OV-101)** in the same hanger.

E. The Intrepid Sea, Air & Space Museum in New York City

Shuttle Enterprise (OV-101)

Stennis Space Center (**SSC**) in Mississippi
The **M**ain **P**ropulsion **T**est **A**rticle, **MPTA-098** without its truss work, is still at NASA's (**SSC**)
<u>NOT accessible, NOT available on display</u>

NASA has several training mock up sections of shuttles but they would need to be pieced together un-naturally to make a complete shuttle.

There is a real possibility that a new Shuttle mock- up might appear one day.

F. California Science Center in Exposition Park, Los Angeles
      2 Shuttles available to seen in the Los Angeles area

Space Shuttle Endeavour (OV-105) and Full-size **Mock up** Shuttle **Downey** AKA *Inspiration (2.0)?*

Full-size **Mock up** Shuttle **Downey**, AKA *Inspiration (2.0)?*

Was Abandon/forgotten by NASA

BONUS

The Downey Shuttle Story

The Original 1972 Space Shuttle Mockup

Abandon/forgotten by NASA

In 1969 while men were walking on the moon. Congress approved the **S**pace **T**ransportation **S**ystem, or **STS** program, later to become known as The Space Shuttle program.

Today we only have a small picture of just how corrupt the Nixon White House was through documents and tape recording. Few know that before Congress approved STS, some contracts were already handed out through bribes to undeserving and or un-qualified companies. Complaints were filed and there were requests for investigations, but that was all lost and held up with Washington's red tape and the Water Gate scandal that ultimately brought down the Nixon Presidency.

One of these early contractors for STS was North American Aviation. Obtaining the new Space contract still was not enough to save the align company. North American had merged with Rockwell-Standard Corporation to become North American Rockwell, Later to become Rockwell International.

North American Aviation and

In July 1972 the North American Rockwell Space Division won the NASA contract to develop and build the new Space Shuttle Orbiter vehicle. Largely due to the full-size mock up built and placed on display at their aircraft manufacturing plant in Downey California. Designed from an original 1962 North American concept for a large orbital transport called the Hypersonic Research Vehicle, The Space Shuttle mock-up was originally built to fulfill a requirement of the contract bidding process. A 122-by-78-foot model largely made of wood and plastic. This shuttle was never officially given a name and was only ever known as the Downey Shuttle.

By 1973 With Rockwell International named the primary contractor for the world's first reusable winged orbiting spaceship: the Space Shuttle. It was here in Downey California that about 12,000 people, stretched over 120 acres in 40 buildings, worked and manufactured the Shuttle parts at the program's peak.

Located in the Design and Engineering Integration Room, or DEI Room, the orbiter mock-up was used for a number of other purposes. It became a centerpiece of public relations displayed on February 27, 1975 and a visual aid for communications with members of congress and other government officials, who frequented the complex of offices and conference spaces across from the mock-up.

Downey remained a useful tool for the program. It lasted for years as an accurately proportioned simulation of the exterior and interior spaces. This mock-up allowed engineers to accurately check prototypes of planned equipment in their intended locations. The Remote Manipulator System (RMS), a jointed, rotating arm made for deploying payloads, built by Canada, is faithfully represented by an accurate, nonworking replica in aluminum. It is bolted to the port side of the payload bay.

Constructed of plywood veneer over a wood frame, and its massive size, this was a formidable prospect. This functioned for years as an accurately proportioned simulation of the exterior and interior spaces of the actual Space Shuttles. The mockup was modified over the years to reflect updates in design. Authentically fitted with every detail, often represented by simple painted wood shapes or silk-screened plastic or metal sheeting. The flight deck, crew quarters, cargo bay, and engine compartment are to size. The mock-up lacks a left wing to eliminate redundancy and to save space.

Mock-ups of payload equipment and containers were also tested for fit, an array of simulated electronic equipment on a specialized pallet that remains bolted in the cargo bay today. Prototype items, such as a machined metal restraint for astronaut's boots at the airlock to the cargo bay remain in place.

By the end of the 70's the public did not care about the Downey shuttle, with real Space Shuttles starting to emerge out of Rockwell's Palmdale facility. Out of sight, it was out of NASA's mind.

NASA had real Space Shuttles to deal with and the Downey Shuttle was not even in their inventory books. Downey was forgotten by NASA.

After the fall of the Soviet bloc Rockwell management started to sell off large divisions of Rockwell International, including its defense and aerospace business. In December 1996 Boeing Integrated Defense Systems bought the former North American Aviation division of Rockwell International. Boeing closed the plant in Downey in 1999.

This left the shuttle and most large items not reusable for Boeings needs abandon.

The city of Downey bought associated buildings of the now-closed Boeing Space Systems Division plant. Plans were for a shopping center, hospital and, movie studio on the 160 acre site.

In 1998, the city started The Downey Studios, renting the facilities for feature film productions by major Hollywood companies, making use of the large indoor spaces, and outdoor open airfields surrounding the many buildings. The studios occupied only a portion of the former plant.

In November 2003 Boeing sold the last of the land, around 60 acres to Industrial Realty Group. At this time Industrial Realty Group took over management of the studios and leased an additional 20 from the City of Downey.

The Downey Studios now featured 79 acres of indoor and outdoor production space including a 50,000 sq ft building, and a 250,000 sq ft building which was home of the largest indoor water tank in North America. A suburban residential street back lot with 5 complete homes and 11 facades was also available at the studio.

In 2003 the Downey Shuttle was forced to move when the space was needed in the main studio building, called Building 1. This was to create an artificial lake for "Lemony Snicket's A Series of Unfortunate Events". (The reservoir served as Lake Lachrymose in the film.) The 65,000 Sq. Ft. Indoor Water Tank, held 6 Million Gallons.

On July 8, 2003, the Downey City council approved a proposal to perform conservation services in connection with its relocation to an area within the building, beyond the perimeter of the movie set being constructed. Construction pressure for the film made the time window to save the shuttle small. The shuttle needed to be documented, taken apart and preserved while moving it a fast as possible. An Omega Nomad data logger was put in place to monitor atmospheric conditions for the Shuttle but was lot with the demolition and construction phase of the movie set. The shuttle as designed, broke apart in to 4 sections for the move. The Original 30 year old casters rolled nicely and the Downey shuttle was placed in its storage location over the next 3 days. A wall was erected and a chain link fence was added to lock away the now preserved Downey Shuttle. With the intention of the Downey Shuttle to eventually become a museum piece a curator was appointed by the city for the next year to regularly check on the secretly stored away Shuttle.

On November 7 2008, a fire damaged part of the studio lot. The fire appeared to have been caused by an exploding propane tank on the back lot. Two left over Apollo space capsules were threatened by the fire which were waiting to be installed at the nearby Columbia Memorial Space Center museum. The Museum would not be open until 2010.

By December of 2011 the locations for all of the Space Shuttles and full size mock-ups had been decided. It was not until June of 2012 that the Downey city council was reminded that they had forgotten their very own Shuttle stored away. With the Downey Studios closed in 2010, Industrial Realty Group wanted to build the Tierra Luna Marketplace. This meant they were going to tear down the buildings and they wanted the forgotten Shuttle moved!

The five-person City Council voted unanimously and approved raising money for moving the shuttle to the Columbia Memorial Space Center museum. They would have to tear down walls and restore the Shuttle in a tent until the new facility was built for the Shuttles home.

On September 29, 2012, The City announced the winning Space Shuttle Mock-Up name contest. The Mock-Up has been named *Inspiration.* They had forgotten to check to see if that name was already taken. The Kennedy Space Center Astronaut Hall of Fame already has a Full size Shuttle mock-up by the same name. It is never too late to change the name. I simply call her *Downey*.

Photo credits:

Hurricane Sandy damage – Jim Henderson

Inside enterprise photos – X plane fan

Smithsonian Discovery - Craigboy

SCA photo - Akradecki

City of Downey

NASA Headquarters Washington, DC

NASA, Dryden

NASA, Kennedy Space Center KSC

NASA, Johnson Space Center JSC

NASA, Marshall Space Flight Center MSFC

NASA, Stennis Space Center SSC

NASA, Jet Propulsion Laboratory JPL

NASA Renee Bouchard

NASA Jim Ross

NASA Tony Landis

NASA Carla Thomas

NASA Tom Tschida

NASA Jet Fabara (USAF)

NASA Tom Reinken

NASA Kevin Rohrer

NASA Bill Ingalls

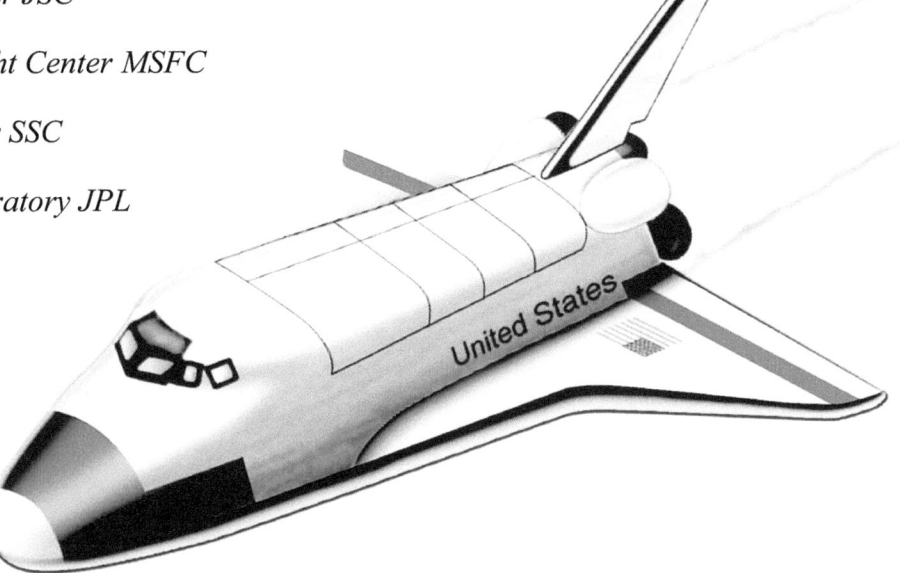

Apologies to anyone missed.

All attempts have been made to clear non-NASA photographers when possible and adhere to the current copy right laws of 2013 and the Fair use act.

www.ingramcontent.com/pod-product-compliance
Lightning Source LLC
Chambersburg PA
CBHW050756180526
45159CB00003B/1484